The Checklist

By Captain Chris Couch

 For information, please contact
Compass Headings Publishing, 9805 NE 116th #7231, Kirkland, WA 98034.

First Edition

ISBN 978-0-9829415-0-8

Write to us:
Compass Headings Publishing, 9805 NE 116th #7231, Kirkland, WA 98034.

captaincouch@earthlink.net

www.CompassHeadings.com

Table of Contents

Maintenance 4
Basic Spare Parts 5
The Lost Horizon 6
Basic Boat Outfit & Required Safety Equipment 7
Troubleshooting 8
Light Show 11
Trip Planning 12
Putting it All Together 13
Overnight Passage & Execution 14/15
Rules of the Road 16
Bar Entrance Crossing 17
Marine Weather Resources 18
Reading the Forecast 19
Weather Tools Online 20, 23-25
She 21
What Does It Matter 22
Where 26
Basic Navigation Tools 27
Terms & Definitions 28
Navigation 28
Marine Radio 29
U.S. Customs 30
Canadian Customs 30
Mexico (and all other countries) 31
Anchoring Tips 32
Using The Ballard Locks 33
Maneuvering Tips 34
Passes and Narrows 34
Seakeeping 34
Distances 35-37
The Presence 39
Directory 40-43, 46-49
Beyond the Horizon 44
Panama Canal 50
Deep Blue 52

Proper maintenance, planning and prudent decision making are the keys to your boating experience. I hope you find this guide useful.

Thank you, ***Captain Chris Couch***

(206) 605-5296 • www.CaptainCouch.com

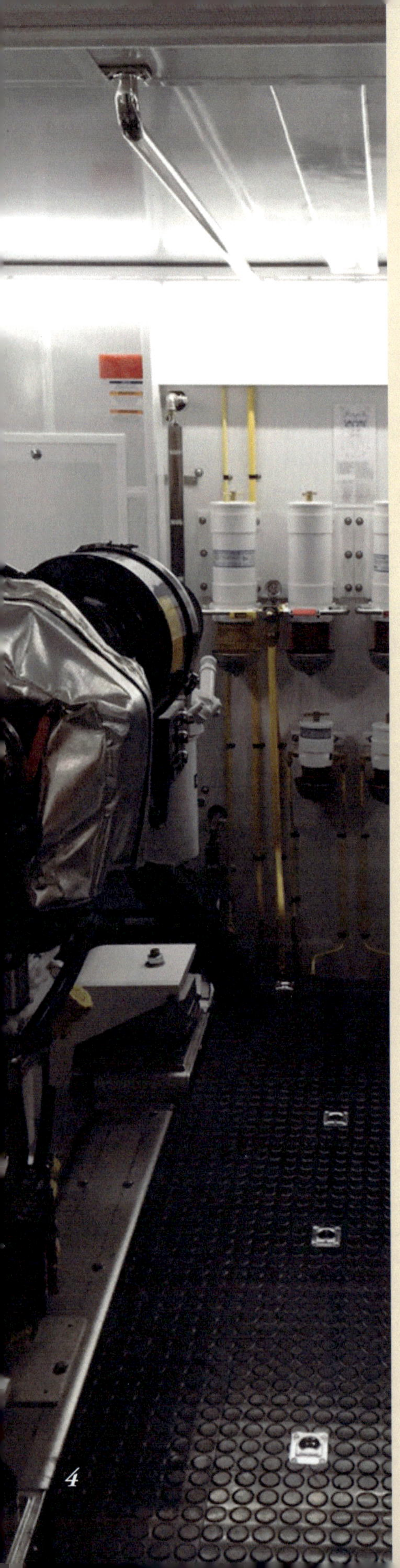

Maintenance

The best maintenance is to use your boat and its systems. Keep your bilges clean and dry. Most electrical problems are due to loose or corroded connections.

Keep a comprehensive Maintenance Log

- ☐ By date for all hull and systems
- ☐ By date and hour meter for main engines and gensets

Don't let these fall through the cracks:

- ☐ Main engine and genset raw water ***impellers***
- ☐ Oil and filter changes
- ☐ Engine and hull zincs - Belts and hoses
- ☐ Thru-hull valves
- ☐ Heat Exchangers
- ☐ Manufacturer's recommended maintenance
  - ☐ Manufacturers scheduled Engine maintenance/checks.
  - ☐ Transmission oil/filter change. ☐ Hydraulic system filter change.
  - ☐ Watermaker High Pressure pump oil change.

Every 12 to 18 Months:

- ☐ Wax exterior/refinish bright work
- ☐ Haul, repaint bottom, replace zincs
- ☐ Main engine/genset checkup by qualified mechanic
- ☐ Check genset valve adjustment
- ☐ Heat Exchanges
- ☐ **All engine zincs**

Every 12 months or every 200 hours:

- ☐ Main engines and gensets
- ☐ Oil, oil filter, fuel filter, racors
- ☐ Water maker high pressure pump oil level
- ☐ Check or replace prefilter
- ☐ Clean/Descale AC raw water intake manifolds

Before each season/before each major trip:

- ☐ Hit all electrical connections with WD-40 or Boeshield.
- ☐ Clean out all shower sumps.
- ☐ Clean out all refrigerators and freezers.
- ☐ Shut down and clean out ice makers.
- ☐ Transmission oil levels
- ☐ Coolant levels. Reservoirs
- ☐ Hydraulic fluid levels
- ☐ Steering fluid levels
- ☐ Hynautic control levels/pressure
- ☐ Check and adjust all flax shaft packings, replace if necessary
- ☐ Sea strainers
- ☐ Engine/genset Impellors
- ☐ Air filters
- ☐ Water filters (replace)
- ☐ Reverse cycle AC/heat ☐ Clean screens ☐ Clean condensation pans
- ☐ Clean bottom and check zincs

Check Monthly:

- ☐ Battery water/connections
- ☐ Start and run main engines
  - ☐ Run thrusters
  - ☐ Run stabilizers
  - ☐ Run tender davit
  - ☐ Run anchor windlass
  - ☐ Exercise steering
- ☐ Excercise Controls, forward and reverse.
  Let run at idle at the dock.
- ☐ Engage and run Aux Come Home System
- ☐ Start and run Gensets ☐ Place under load
- ☐ Turn on and operate all electrical
  - ☐ Equipment, appliances and electronics
- ☐ Turn on and check navigation lights
- ☐ Run water system
  - ☐ Keep water in tank fresh
  - ☐ Replace filter annually
- ☐ Run bilge pumps and float switches
  - ☐ Keep bilges clean and dry
- ☐ Exercise all thru hull valves
- ☐ Flush all heads. Keep holding tank pumped out
- ☐ Run Watermaker (every 2 weeks)
- ☐ Pickle if not being used
- ☐ Main engine shaft packings
- ☐ Rudder shaft packings

Refeuling:

- ☐ Add Biocide each time

Tender:

- ☐ Service and run outboard prior to season
- ☐ Clean and ensure tender is properly equipped
- ☐ Flush outboard after salt water use
- ☐ Battery switch to "off" after each use
- ☐ Check bilge pump and float switch

Sailing Vessels:

Every 12 months / Before each Season

- ☐ Check and Run Roller Furling Systems
- ☐ Check and adjust tension on Steering system cables
- ☐ Check all Turnbuckles for corrosion
- ☐ Check all Stays and Shrouds
- ☐ Have Rigging tuned
- ☐ Service Winches
- ☐ Lube and check Pawls

Monthly:

- ☐ Operate all Winches
- ☐ Operate Roller Furling
- ☐ Air out and dry sails.

Basic Spare Parts

Look at each system onboard and identify the components that are prone to failure. Determine the operational and or safety importance of that system.

- ☐ **Main Engines/Gensets:** Oil filter, fuel filter, racors, alternator belts, impellors, zincs. Fuses for Genset control box. Contact engine dealer for engine spare parts kits.
- ☐ **Fluids:** Coolant, engine oil, transmission oil, hydraulic fluids. Identify all hydraulic systems. Steering fluid (i.e. Hynautic, Sea Star)
- ☐ **Fresh Water System:** If no auxiliary pump is installed. Spare pump/motor and pressure switch. Filters.
- ☐ **Water Maker:** Primary and secondary filters. Pump oil.
- ☐ **Toilet/Holding Tank System:** Components particular to system. Macerator pump.
- ☐ **Navigational Lights:** Spare bulb for each type.
- ☐ **Electrical:** Identify breakers, fuses and switches that may be needed.
- ☐ **Bilge Pumps:** Spare pump and switches.
- ☐ **Batteries:** Distilled water.
- ☐ **Small Boat:** Oil/fuel filters, oil, spark plugs, flush attachment.
- ☐ **Tools:** Basic tools, filter wrench. electrical repair. Rags.

Notes:

The Lost Horizon

by Chris Couch

Standing at the edge
Where my faith and doubt collide
Just as time starts to stand still
How do I keep moving forward?
And step to the other side

Because it has never been very clear
Just how deep I can go?
Is there another way?
Or is this the only world
I know?

What is this feeling?
Of which I am unable to say
How do I keep it?
Before
It is taken away

Years spent searching these shores
What has all of this meant?
Perhaps, if I knew the reason
For which I was sent

I know I was changed forever
By a place I have never been
I touched it in a dream
I just don't remember when

I am not sure

Am I still searching?
Or do I just need to "be"
At this Lost Horizon

That is really me

Basic Boat Outfit & Required Safety Equipment

Deck

Mooring lines
Fenders
Telescoping handle
Soft bristle brush attachment
Boat Hook attachment
Chamois mop attachment
Chamois, microfiber
Large plastic bucket
Washdown soap
Water hose w/nozzles
Plexis pexiglass cleaner
Cleaner wax
Acetone
Teak deck cleaner

Tools

Tool box
Basic sockets/wrenchs/drivers
Assorted crescent wrenches
Filter wrench
Waterfilter wrench
Very large channel locks
Vice grips
Hammer
Hack saw
Tape measure
Large divers knife
Hearing Protection

Electrical

Small tool box
Multi-meter
Wire stripper/crimper tool
Small wire cutters
Needle nose pliers
Bronze wire brush
Assorted connectors
Assorted tie wraps
Electrical tape
WD-40

Misc. Repair

Large bag of rags
Oil sorbent pads
Wet/dry vac
Assorted oil funnels
Air pump (for bicycle tires)
Silicone sealant
5200 Marine adhesive
Masking tape, 3M blue
Two-part Epoxy adhesive

Operation, Navigation and Safety

Life jackets
Throwable ring
Fire extinguishers
Bell
Flares, three day/night
Rules of the road (Colregs)
Oil and garbage placards
Flashlights/batteries
Binoculars
Paper charts
Guide books
Dividers/parallel rule
Notebook/pens/pencils
Scissors
Easy liner rubber matting
Assorted bungee cords

House Keeping

Vacuum cleaner
Glass cleaner
Dish soap
Soft scrub
Paper towels
Toliet paper
Toliet bowl brush
Toliet bowl cleaner
Plastic containers with lid for parts stowage

Tender

Tie-down straps
Lines/fenders
Small anchor/line
Oars
Motor flush attachment
Lifting cables
Flares
VHF-FM radio

Recommended Equipment

Marine VHF radio
Adequate bilge pumps
Anchor and line
First aid kit
Flashlights
Boat hook
EPIRB
I.D. for each person onboard
Annual customs sticker
Crew list
Boat insurance
Import license
Fishing licenses
Mexico paperwork.

To obtain Mexico paper-work: Romeros Mexico Service 949-548-8931

Required

- [ ] Certificate of document Onboard with annual sticker
- [ ] Name/home port & official number properly displayed
- [ ] State registered current registration onboard
- [ ] Tender registration current registration onboard
- [ ] Numbers properly displayed
- [ ] Personal floatation devices. One for each person and proper type
- [ ] Throwable floatation device. At least one
- [ ] Fire extinguishers Proper number and type for size of boat. Inspected annualy or in serviceable condition.
- [ ] Visual distress signals 3 day/night. Not expired.
- [ ] Sound producing devices. Proper size Bell for size of vessel. Does not have to be installed
- [ ] Rules of the road. Copy onboard.
- [ ] Pollution and trash placards. Properly displayed.
- [ ] Marine sanitation device Proper type.
- [ ] Navigation lights
- [ ] Ventilation - gas engines

Troubleshooting

Basic rules of thumb:

- Keep bilges clean and dry.
- Understanding how a piece of equipment or system functions is key to troubleshooting its malfunction.
- Most electrical problems are caused by bad connections.
- Many problems can be solved by turning unit off and then power to the unit at the breaker. In essence doing a reset.
- Smell can indicate problems. Coolant, overheated batteries, fuel system, black water all have definite smells.
- Look, Listen, Smell, Feel.

Trouble Underway:

- Maintain control of your vessel.
- Place vessel on safe/stable heading.
- Control the damage. Fire, flooding, fluid leaks, etc.
- Get boat in safe/stable condition.
- Call for CG assistance if necessary. Ch-16.
- Head for nearest safe port.
- Do not attempt permanent repairs underway.

Bilge Pumps:

- If runs but does not pump. Maybe air bound. Cycle on and off.
- Check back flow valves. Maybe stuck.
- Check power.
- Check Manual and float switch operation.
- Check for debris that may clog intake.

Fluid in the Bilges:

Water in the bilge:

- Fresh or Salt? Taste test. Possible sources?
  Fresh: Fresh water system, deck drain system
  Salt: Thru hull, raw water cooling, shaft packings
- Coolant: Will have definite color and smell.
- Fuel: Typically valves and fittings. Color and smell.

Oil: Clean or dirty: Trace back to source.

- Clean: Transmission. Hydraulic. Steering.
- Dirty: Engine or genset
- Hydraulic oil: Trace back to source: High pressure lines and fittings are typical sources. Look for chaffed lines. Color of oil may also indicate source.

Stabilizers, Thrusters, Anchor windlass, Boat Davit.
Steering system may have different color fluid. Ie. Hynautic

Hydraulic Systems:

- Check power and selonoid switches.
- Check high pressure relief valves. May be stuck.
- Check filter. May be clogged.
- Check fluid level.
- Check pump output. Pressure gauge.
- Check valve bodies. Operate manually.

Electrical:

Basic:

- Cycle on/off switch.
- Check breaker at the panel.
- Check power connections.
- Check ground fault outlets. May have tripped.
- Check for power supply to the panel.

Battery Systems: House, Start

- Check main DC breaker at main panel.
- Check engine battery selector switches.
- Check DC voltage.
- Check main battery switches.
- Check battery connections.
- Check battery chargers.
- Underway: Check alternator output. It is recommended that one engine Alternator charge the house bank and the other engine alternator charge the start bank.

Inverter:

Won't Charge:

- Check inverter breaker.
- Check AC in.
- Cycle unit on and off. Cycle charge on and off.

Won't Invert:

- Cycle invert switch on and off.
- Check main DC power to unit.
- Check power selector switch on main panel if present.
- Check for proper House Bank Voltage.

Shore Power:

- Check boat's main panel. Ensure shore power is properly selected.
- Check dock breaker
- Boats main breaker (usually located near shore cord entry)
- Test dock plug, test/inspect shore cord. Maybe loose connections in the plug.
- If using a Smart Y to make 240VAC, each 120/35 amp supply used must come from a separate shore feed. (different pedestals). Remember, you will only have 35 amps, so manage your usage.

Octoplex

- Reset entire system
- Reset Control screens using end of a paperclip
- Check Inverter operation. Some circuits may be powered by the Inverter

Troubleshooting

Genset Power:

- Check that Genset is properly selected on main panel.
- Check main breaker on Genset.

Genset shutdown:

- Most Gensets will shut down with loss of oil pressure
- Overheating or loss of raw water flow.
- Overheat: Impellor, raw water supply, sea strainer, coolant.
- Is impellor pump cool or hot.
- Oil Pressure: Oil level, bad sending unit.
- Raw water supply: Sea strainer. May be clogged.
- Check for coolant leak.
- Shutdowns maybe malfunctioning.
- Check heat exchange for blockage and corrosion.
- Check GenPak control box fuses.
- If your gensets are connected to the Fire boy system, this could be the problem.

Main Engines:

Won't start:

- Check that transmissions are in neutral.
- Check that electronic controls are selected.
- Check ignition power breakers/switches at panel and engine room.
- Check power connections at starter.
- Check remote start/stop on engines. (Cat)
- Check battery connections.
- Check ignition switch operation.
- Check Start Battery and Control System Voltage.

Main engine overheat:

- Check sea strainer.
- Check raw water impellor.
- Check heat exchanger coolant level.
- Check heat exchange for blockage and corrosion.

Main engine loss of RPM or power:

- Shut down immediately to avoid running
- Engine out of fuel.
- Shift to second racor if present.
- Check for proper fuel supply valve alignment.
- Check fuel tank level.
- Check suction gauge on racor.
- Visibly check racor.
- Check air filter and intake.

Electronic Engine Shutdown:

- Loss of DC voltage will shut engines down.
- Check ignition switch.
- Check battery connections.
- If your engines are connected to the Fire boy system, this could be causing the problem.

Fuel System

Understand how fuel flows from the tank, to the Racor Filter, and then to the engine. Understand your Fuel Manifold and Management System.

- Check bowl of racor filters for water.
- Always keep plenty of spare filters onboard.
- Always put biocide in when refueling.
- If equipped with duel racors, use one at a time.
- If equipped with sight guages, keep bottom valve closed.
- Replace sight gauge tubing if faded and cloudy.
- Always compare fuel gauges with actual fuel used.

Watermaker:

- Valve alignment, pressure setting/valve, filters, salinity level
- High pressure pump oil level.

Fresh Water System:

- Turn Power off and back on to Pump. Hit reset if present.
- Check water tank level.
- Check valve alignment.
- Check backup pump.
- Check Accumulator Tank pressure. Should be set at pressure switch kick on pressure.
- Check pressure switch.
- Some pressure switches will not kick on with
- Zero pressure in the system.

Head Systems:

Vacuflush: If it won't stop running or cycling. Flush several times running lots of water through the system. The rubber duckbill valves may have material stuck in them. Check for air leaks in the system. Check the pressure switch on the accumulator tank.

Headhunter, Techma and Planus

Check power, All of these have a momentary switch,
Water supply valves, Water supply solenoid operated valves.
The timing adjustment for these is located either at the switch or on a control box near the solenoid.

Troubleshooting

Holding Tank Discharge:

- Ensure overboard discharge valve is open.
- Diaphram pumps: Rubber Duck bill valves maybe turned inside out.
- Do not run with overboard discharge valve closed!

Steering

- Head boat down swell and steer with controls.
- Check Lazarette for mechanical linkage failure or loss of hydraulics.
- Check Hydraulic Resevoire level and pressure.
- Check power assist pump. Usually located on front of one of the main engines.

Air Conditioning:

- Control head operation.
- Ensure system is selected to cool or heat.
- Raw water cooling pump.
- Raw water sea strainer.
- Air handler screens. Check if dirty.
- Compressor may need recharging.
- Check circulation water level and pressure.
- Circulation water pump.
- Check for proper Shore power voltage.
- Manifolds may need to be cleaned/descaled.

Electronics:

Radar:

- Check proper adjustment of: Gain, Sea Return, Rain, Tune.
- Adjust all for best picture.
  Note: Any one of these can eliminate the picture.
  Check for accurate alignment of the radar picture.

Depth Sounder:

- Check gain adjustment.
- High frequency for shallow water.
- Low frequency for deep water.

VHF-FM Marine Radio:

- Check for TX indicator when Mike is keyed.
- Test with other radio if possible.
- Check Antenna connections.

AutoPilot:

- Check or adjust rudder settings. If the pilot has trouble holding a course or struggles in a following sea, the settings need to be increased.
- Check rudder feedback unit next to steering gear.
- Compass may not be properly swung.
- Reset to Factory settings if all else fails.

Stabilizers:

- Adjust Gain or Sensitivity.
- Adjust Trim control if present.
- Check settings. Reset to Factory settings if needed.
- Check for loss of Hydraulic pressure. (Hydraulic Systems)

Vibration:

Possible causes

- Engine and shaft coupling alignment
- The shaft itself
- Strut bearings
- Prop balance
  Note: Bad injectors can also cause roughness or vibration.

Light Show

by Chris Couch

Rocking gently
In a following sea
A warm moist breeze
Blows gently past me

Looking back
As my world slowly spins
This day is setting
Somewhere, a new one begins

My mind fills quickly
with dreams full of stars
How do you describe
This place in which we are

A spirits glittered trail
Brightens the truth
Against the darkened skies
Oh, the things I have seen
With these eyes

Alone I stand
Sound of the water
The smell of the sea
Her light show
That surrounds me

This moment washes softly
Across my face
As her thoughts wet my lips
It is the sea that I taste

Trip Planning

Proper maintenance, planning, preparation and prudent decision making are the keys to your successful boating experience.

Pre-Trip

- ☐ Ensure all maintenance and necessary repairs are completed and up-to-date
- ☐ Review maintenance checklist
- ☐ Check safety equipment and required documentation
- ☐ Safety and required equipment checklist
- ☐ Review route to be taken. Make note of waypoints, aids to navigation and landmarks to be used
- ☐ Ensure route taken is in safe water
- ☐ Ensure safe water between waypoints
- ☐ Stay in marked channels
- ☐ Do not cut corners
- ☐ Call destination Marinas

What will be your:

- ☐ Time of departure
- ☐ Time en route
- ☐ Time of arrival
- ☐ Fuel onboard
- ☐ Fuel needed

What entrances, passes, narrows, channels will be transited and at what times?

Bar Entrance Crossing/Passes and Narrows

- ☐ Look at tides and currents that may apply
- ☐ Call enroute to destination marinas and fuel docks
- ☐ Database

Look at Forecast Weather

- ☐ Are forecast winds and seas within your safety limits
- ☐ Weather Resources

Departure

- ☐ Ensure boat is ready and properly secured: port lights, hatches, loose items, etc.
- ☐ Ensure each passenger is ready. Proper clothing, ID, medication, etc.
- ☐ Check current and forecast weather
- ☐ Make go/no go decision. The most important decision you will make!
- ☐ Review your en route trip planning
- ☐ Start Genset. Bring in all shore ties
- ☐ Start electronics
- ☐ Start engines. Test controls
- ☐ Determine onsite wind and current to properly and safely depart dock
- ☐ Always think ahead. Plan ahead

Arrival

- ☐ Call ahead and determine what slip or dock you will be at
- ☐ Look at onsite wind and current How will that affect your maneuvering
- ☐ Set up lines and fenders before entering the marina
- ☐ Remember: It is easier to back into the wind and current

Putting this all Together

Lets bring everything together into a trip plan.

Departure point… Arrival point.

What is the enroute distance?

What speed will you run?

What will be your time enroute?

Are you making an overnight leg?
Ensure the overnight portion is beyond Crab Pot areas.

Are you departing and or arriving at a Port with a Bar Entrance?

Ensure departure and Arrival is to be in the Daylight.

Ensure departure and Arrival is done at a Flood or Slack Tide.

Do you have a safe Weather Window for your leg or trip?

- What is the Departure Forecast?
- What is the Enroute Forecast?
- What is your Arrival Forecast?

The Overnight Passage

To the uninitiated, doing an overnight passage offshore can be a very intimidating experience.

Here are some basic rules to help make it a better experience.

Pick really good weather. Light winds and light seas.

Ensure everything in the boat is properly secured.

Use towels, tape, ext to darken the pilot house and console as much as possible. This is very important. All lights and displays should be dimmed or covered.

Keep flashlights handy.

Have everyone camp out in the Salon if possible.

Ensure that your track takes you well outside crab pot areas. Deeper than 400 ft.

Execution

A trip up the inside passage to Alaska or down the coast to Southern California or Mexico is nothing more than a series of legs.

Each leg is executed separately.

You know your departure time based on daylight and proper tidal current.

You know your arrival time based on daylight and proper tidal current

You know your route, distance, speed and time enroute.

You have picked a good weather window for this leg.

Your vessel and crew are ready.

You have made a last minute check of realtime weather and latest forecast.

Now safely execute this leg.

Rules of the Road

The International Rules of the Road were formalized by the convention on the International Regulations for Preventing Collisions at Sea in 1972. They are commonly called the 72 COLREGS. The Inland Navigation Rules were enacted by law in December of 1981. These regulations govern actions and responsibilities between vessels. If you are a licensed mariner you should be versed in these rules and regulations. If you are NOT familiar with the Rules of the Road, then the following common sense tips should help.

Drive Defensively

- Stay clear of all commercial traffic.
- Power-driven vessels will stay clear of vessels under sail.
- Assume that all other boaters are ignorant of the Rules of the Road.
- Drive defensively. Stay clear of all other traffic. Use common sense and courtesy and err on the side of safety when making decisions.

Between You and the Other Vessel

- When in a marked channel or fairway, stay as far to the right as possible.
- Determine Risk of Collision with crossing traffic as soon as possible. (Determine Relative Bearing Drift)
- When you make a course change, make it large enough so that there is no question as to your intentions.

Crossing

When you have the other vessel on your right side, it is your responsibility to stay clear or pass to the stern of that vessel.

Overtaking

When you are overtaking or passing another vessel it is your responsibility to stay clear of that vessel.

Meeting Head On

When meeting another vessel head on, when possible alter course to the right. In any case make your course change large enough so that there is no question as to your intentions.

Risk of Collision

- Determine Relative Bearing Drift.
- When crossing traffic is spotted, note its relative position using a reference point on y our vessel. Using that same reference point, note the relative change left or right. If the relative bearing of the other vessel does not change, then a risk of collision is present.
- Take action to allow the vessel to pass either ahead or behind you with a wide margin of safety.

Vessel Wake

- Your wake is your responsibility. Use courtesy and common sense with regards to the wake your vessel is producing.
- Always be aware of how your wake is affecting other boats, docks, structures etc.

The Bar Entrance

The following entrances are considered Bar Entrances and can be hazardous especially during an Ebb Tide: West Entrance Strait of Juan De Fuca (can be hazardous with an incoming sea or swell and ebb tide), La Push, Grays Harbor, Columbia River, Tillamook Bay, Depoe Bay, Yaquina Bay, Siuslaw River, Umpqua River, Coos Bay, Brookings OR, Humboldt Bay, Fort Bragg, San Francisco, Morrow Bay.

Bar entrances can become hazardous when you have an incoming chop or swell meeting the outgoing Ebb Tide. The outgoing current slows and builds the incoming chop or swell. The flood current will actually flatten the incoming wave.

I recommend crossing only on the Flood or Slack. Call the Coast Guard on channel 16 if in doubt. During certain conditions, bar restrictions may be in place.

The following entrances are "NOT" Bar Entrances and can be transited at any time: Crescent City, Bodega Bay, Half Moon Bay, Santa Cruz, Monterey, any port from Point Conception south.

Crab Pot Territory

Crab pots will be found along the entire West Coast down to Southern California and then Lobster Pots from there south into Mexico. They are the most numerous around port entrances from Cape Flattery to San Francisco.

My general rules are: Arrive and depart from any entrance only in the Daylight. At night, stay in water at least 400 ft or greater.

The only entrance along the West Coast that is safe to approach at night is Monterey.

Daylight Only

Because of debris and pots in the water, I recommend transiting the following areas "only in the Daylight". This applies to Mexico as well.

- All of the Inside Passage from Puget Sound to Alaska.
- All inland waters inside of Cape Flattery.
- The Columbia River.
- Any Coastal Waters less than 400 ft.

Note: Because of Crab pots, the only entrance on the West Coast that can be safely approached at "Night" is Monterey.

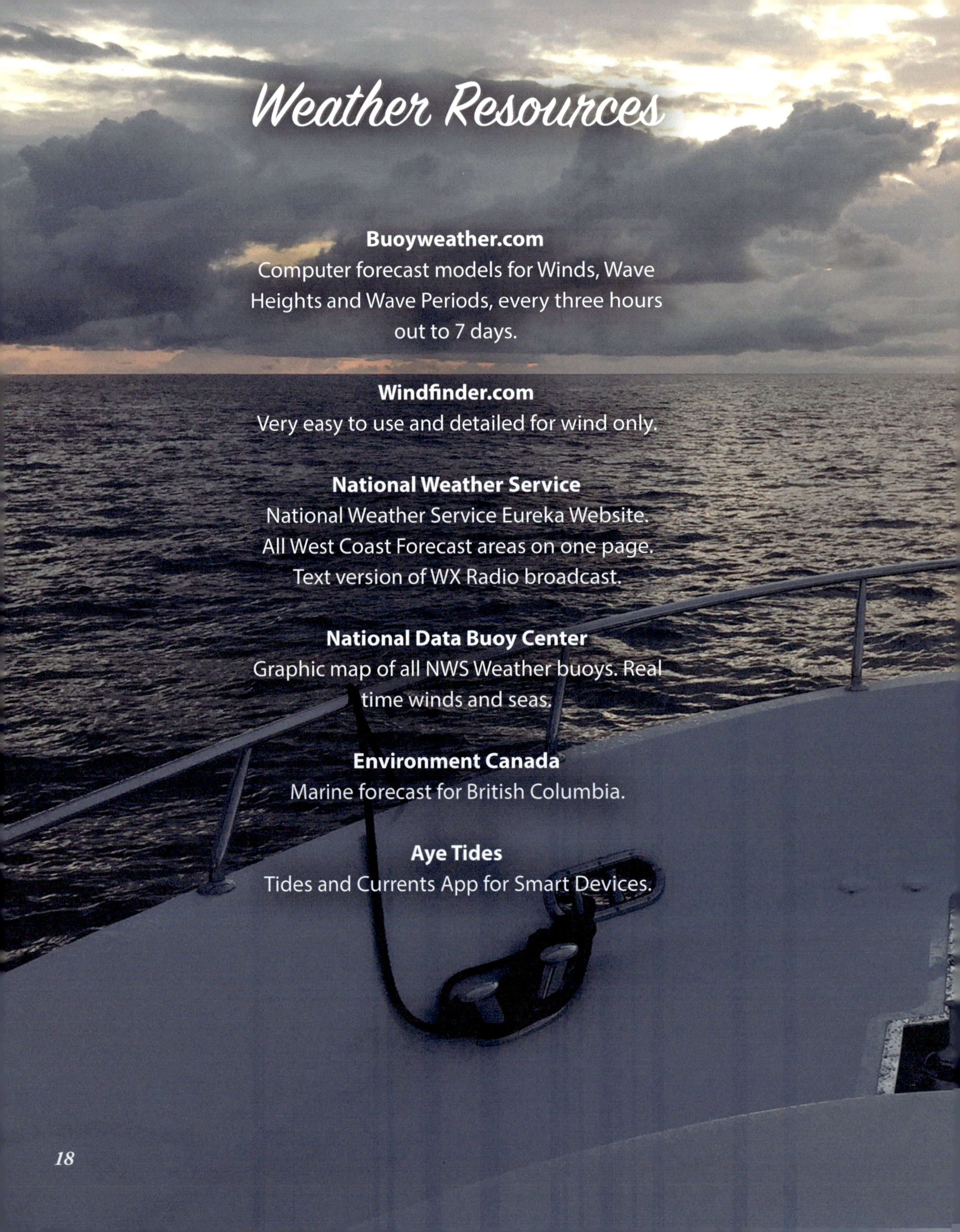

Weather Resources

Buoyweather.com

Computer forecast models for Winds, Wave Heights and Wave Periods, every three hours out to 7 days.

Windfinder.com

Very easy to use and detailed for wind only.

National Weather Service

National Weather Service Eureka Website. All West Coast Forecast areas on one page. Text version of WX Radio broadcast.

National Data Buoy Center

Graphic map of all NWS Weather buoys. Real time winds and seas.

Environment Canada

Marine forecast for British Columbia.

Aye Tides

Tides and Currents App for Smart Devices.

Reading the Forecast Maps

All of the weather websites get their information from the same place. The North American Model.

But, they all portrait this information in different ways.

Once you have found the site that suits you the best and you have become familiar with the way that site operates, here are some basic rules.

Adjust the map and zoom in to show the area you are transiting at the best scale.

Look ahead each day until you find a suitable weather window. Winds, seas and periods within your limits.

Watch that window for several days thereafter to ensure that the forecast models are "consistent".

Watching for consistency in the models over several days is extremely important. If the models keep changing, then you are unable to be sure of what you may actually see.

Note: Computer Models will not forecast inland areas and will not forecast Cape Effect phenomena.

Weather Tools

Buoyweather.com

Buoyweather.com is a subscription site that provides computer forecast models for Winds, Wave Heights and Wave Periods. Maps can be viewed for every three hours out to seven days. There are many weather sites that do basically the same thing. This is the one that I have used the longest and trust the most.

Getting started: Navigating to the forecast maps page:
From the home page:

- Click 16 Day Forecast
  - Premium Charts and Forecasts
    - Click on Regional Charts
      - Click on North East Pacific or the area you are interested in
        - Save into favorites

- Use Current Location to change the center of the map and then to zoom in to the area you want. Use Wave Height, WW3 Wind and Wave Period.

This will now be in your Favorites every time you log in.
Use the menu below the map to view WW3 Wind, Wave Height and Wave Period.

She

by Chris Couch

Before me she lays
Old and wise
Her ancient hues
Reflect from my eyes

Sun shines when she laughs
Rain falls as tears
What is the hold she has had on me
All of these years

Humility has been taught
Respect she demands
Patience has been learned
I will obey her commands

I sail in her presence
She allows me in her world
I walk her laced edges
Her swells gently curled

Why am I drawn to her
This place from where we began
This force that humbles me so
I don´t know
I just am

I have known her fury
I have seen her wrath
Her beauty moves me

She is my path.

What Does It Mean?

By Chris Couch

What does it mean
To question
The essence
Of who
We really are?
Do you wish
Upon
A falling star?

Why do we choose
The fear
In which we live
And the lies
We know?
What does it
Feel like
To finally
Let go?

What is the truth
I touch
As I sail
These open seas?
What does
It mean
When it brings me
To my knees

The darkness
Of my journeys
Night
What is the dream
That gets me through?
What does it mean?
Missing you
The way
That I do

What does it look like
This life
That I see?
Will you take
My hand
And take this journey
With me?

I have held beauty
And have watched it
Die
Stepped
To the other side
And watched
The angels cry

As my journey
Flows through
Her eyes
Am I able to be?
Just
As wise

I ask of her

I know who I am
But I don't know
From where

She's answers
Because

You are already
There

1.

National Weather Service Eureka

www.wrh.noaa.gov/eka

NWS Eureka's website is the easiest to use especially for mobile devices. It gives you all of the West Coast Forecast Areas in one place.

From this page:

1. ➤ Click on Marine Weather, on the left under forecasts

2. ➤ Click on Other Marine Links tab in the upper right

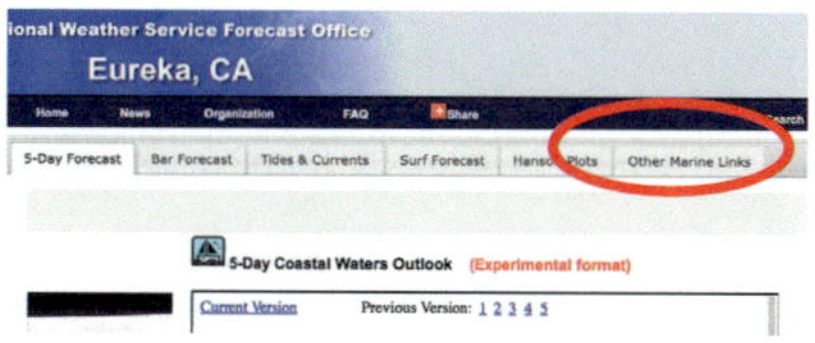

2.

This page has every forecast area for the West Coast.

For Puget Sound and the Strait of Juan de Fuca, ➤ *Click on Cape Shoalwater to Cape Flattery and scroll down.*

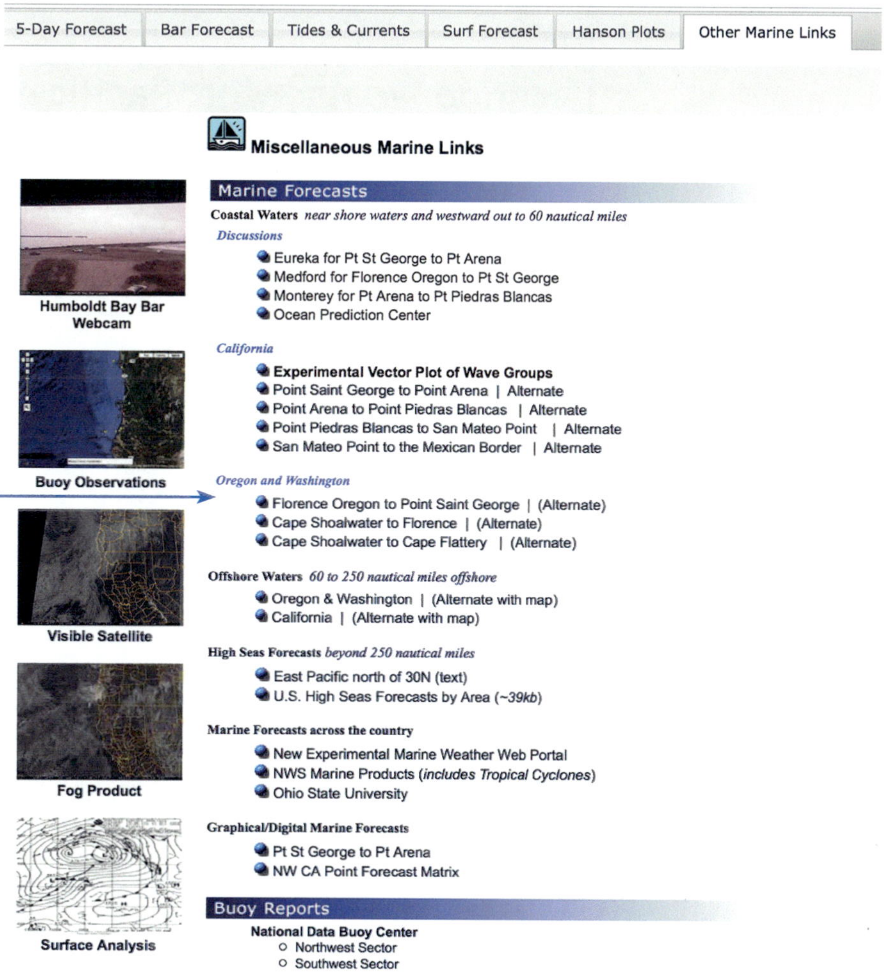

Environment Canada

weatheroffice.ec.gc.ca
Environment Canada is the Canadian version of the National Weather Service.

To navigate to the Marine waters of British Columbia:
From the Home Page.
➤ Click on Marine
Under Marine Forecasts, ➤Click on Pacific - Georgia Basin
➤Click on the area of interest.

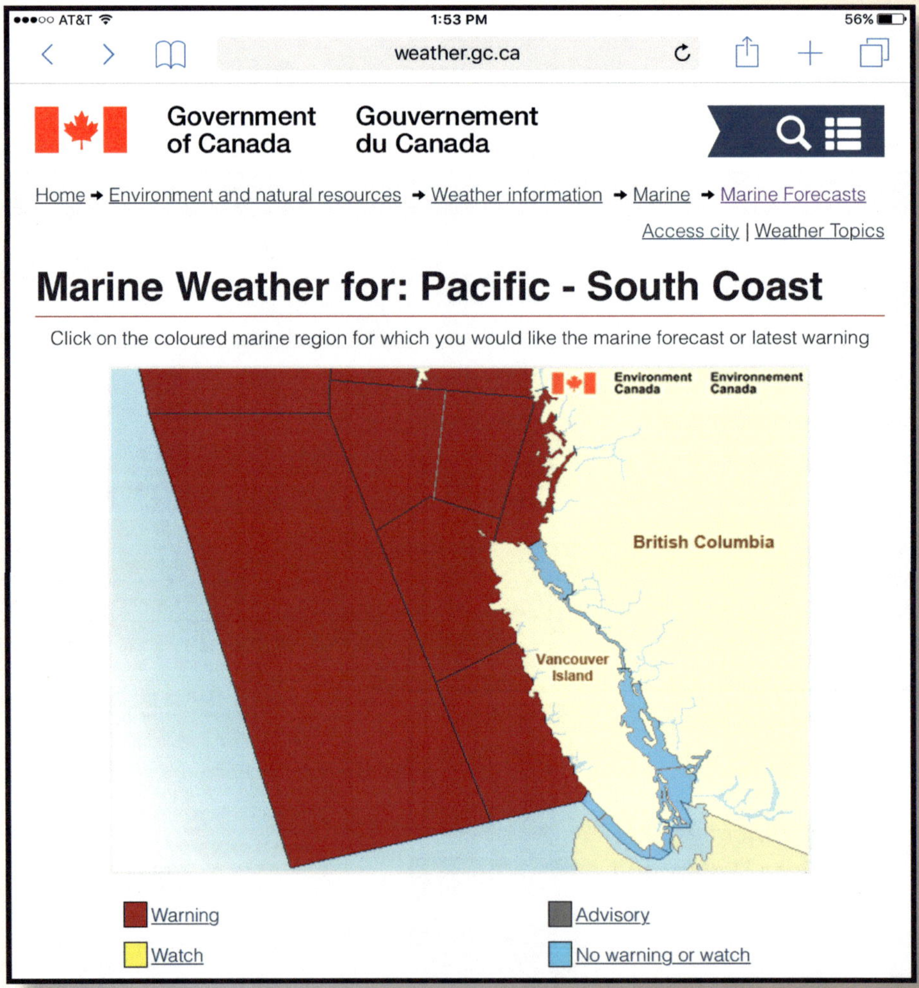

National Data Buoy Center

ndbc.noaa.gov

National Data Buoy Center provides an interactive map with the locations of all weather buoys.

This gives you real time winds and seas for each buoy location.

All marine areas are subdivided into regions. By clicking on the region and then the buoy of interest, you will get real time observations for winds, seas and periods.

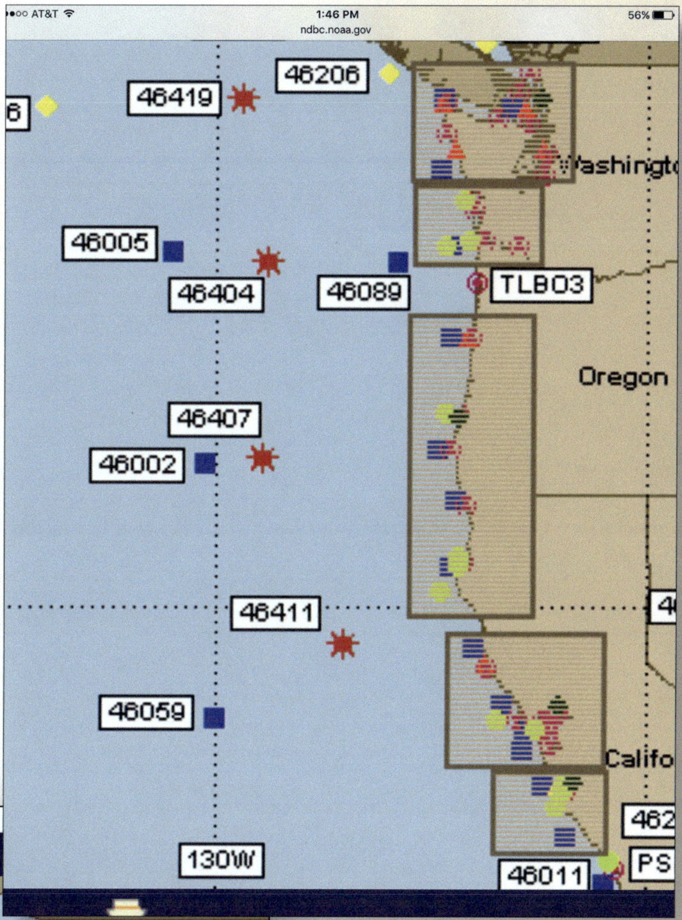

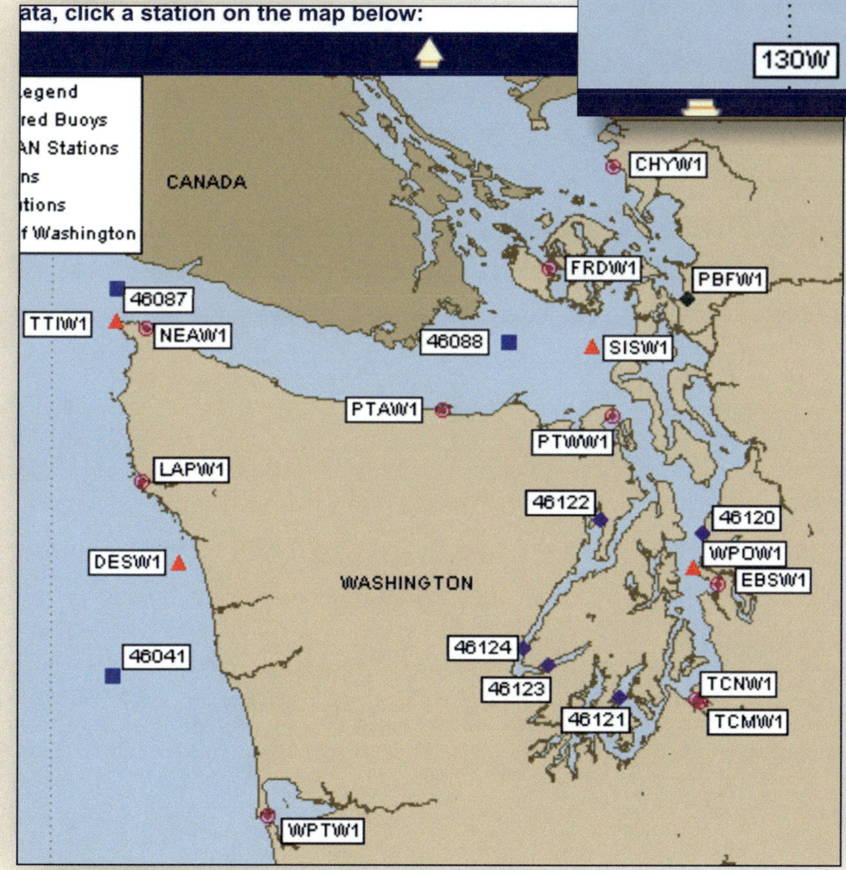

Where

by Chris Couch

Where do I go
When I sail off to shores
Never been?
How will I get there?
As she rocks me gently
Once again

Why have I sacrificed
So much to be alone?
In this place that feels
So much
Like home

Everytime
She reaches into me
How do I learn to listen
How do I learn to see

The visions she shows
With each swell
That rolls by
I know I belong
But I don't know why

What is wrong?
And who is right?
Why am I here?
I ask of her
Each night

Where it all started
Life that blows in the wind
Where will I go?
When this all
Comes to an end

Basic Navigation Tools

Always be in the habit of correlating what you see visually with what is on the chart and with what is displayed on the radar. Buoys, lights, points of land, shoreline, other boats.

Note: The more you practice this, the better you will become and the safer you will be. Cross-check what you see outside with what you see inside. Do not stare at your plotter!!

Visual

From what I observe, far too much time is spent focused inside the pilothouse staring at the chart plotter. Look outside, look ahead, identify what you see outside with what you see on the chart.

Chart Plotter

Like Visual, Radar and your Depth Sounder, your Chart Plotter is a tool. Learn to Cross Reference it with your other tools. Do not completely rely on it!

The Radar

Your Radar transmits on a frequency in the Microwave band. It is line of sight. Proper adjustment is the key to your Radar's operation. Check proper adjustment of: Gain, Sea Return, Rain and Tune.

Note: Any one of these can eliminate the picture.
Adjust each one for best picture. Check for accurate alignment of the radar picture.

Aids to Navigation

Lights, Buoys and marks are there to mark the navigable channel and hazards or obstructions to navigation. Never assume on what side to take a light or buoy. Always check your chart!! It is there for a reason!!

The Depth Sounder

Is your sounder reading depth under the keel or under the transducer. There can be a difference of six inches to a couple of feet. Ensure that your offset correction is properly entered. Your depth sounder can be just as important as your chart plotter. Always be in the habit of checking depth against the chart.

Ensure your sounder is properly adjusted.

Check gain adjustment: high frequency for shallow water, low frequency for deep water.

The Magnetic Compass

Magnetic Compasses are inherently inaccurate. You can have five compasses onboard and they will all read something different. Compasses are affected by anything metal, electric and electronic.

The Autopilot

Your Autopilot is a very useful tool. It handles the very tedious chore of steering the boat while freeing you up to focus on safely operating and navigating your vessel.

Your Autopilot uses a Magnetic Compass. When properly set up, the heading you select will be relatively close to your desired course over the ground, but not exact.

Operation

If the pilot has trouble holding a course or struggles in a following sea, the settings need to be increased. Go into setup and check rudder settings. They may be too low.

Check rudder feedback unit next to steering gear.

Compass may not be properly swung.

Terms & Definitions

Vessels Heading (HDG)
Stand at the Helm and sight down the bow of your vessel. Look at what your Magnetic Compass reads. This is the Vessels Heading.

Range (RNG)
Distance to Waypoint. Should be in Nautical Miles.

Course over the Ground (COG)
Your GPS takes several fixes per second and plots them on a digital chart. The direction of the track created by this succession of fixes is your Course over the Ground.

Note: Depending on the vessels drift created by wind and current, your vessels heading may be several degrees different. Do not confuse the two.

Speed over the Ground (SOG)
The same succession of fixes plotted by your GPS is used to calculate your Speed over the Ground.

Note: Speed over the Ground will change with the direction and speed of the water current.

Bearing to Waypoint (BNG) Or Course to Waypoint (CSE) The direction from your vessel to the waypoint as referenced on the Compass is your Bearing to the Waypoint.

Navigation

Planning Your route
The route to your destination will be a series of legs.

Make a note of each leg to be done. Review navigational references to be used: Buoys/Lights, Points of Land, Location of Waypoints to be used.

Consider hazards to navigaton currents, weather and aids to navigation in the planning of your route.

The Waypoint
Ensure that there is always safe water between your current position and the Waypoint you have created.

Navigating to your Waypoint
Ideally, you want the Course Over the Ground (COG) to match the Bearing to Waypoint (BNG). When you keep the two matched, your vessel will follow the track line created by your Chart Plotter.

Select a vessel heading that gives you the desired

Course over the Ground.
Course over the Ground (COG) vs. Vessels Heading (HDG) vs. Bearing to the Waypoint (BNG) or (CSE), or sometimes referred to as Course to the Waypoint.

It is very important to understand the differences between these terms.

All will be described as a direction referenced to the

Magnetic or True Compass. Zero degrees (North) 90 degrees (East), 180 degrees (South), 270 degrees (West), back around through 359 degrees.

Most boats will only have Magnetic. It is recommended that everything be kept in magnetic to avoid confusion.

Only vessels equipped with special GPS receivers or a Gyro Compass will have the ability to reference True.

Marine Radio

VHF-FM Marine Radio

Your VHF-FM Radio is a valuable tool, not a toy.

Do not use it for BS conversations.

It is to be used to call the Coast Guard,
other vessels for navigational safety and marinas.
Keep your radio tuned to channel 16 at all times.
When possible transmit on one watt to keep your transmit range low.

Calling another Station

Example: "US Coast Guard this is (your vessels name) Channel 16 over."
"(Other Vessel) this is (your vessels name) Channel 16/13 over."
Marina XYZ this is (your vessels name) Channel 68 (or their designated channel) over.
When making a distress call to the CG, ensure you give your position first. Always know where to find your Latitude and Longitude and know your position by geographic reference.

Mayday Mayday

Use only when in immediate distress. Channel 16
Example:
Mayday Mayday this is (your vessels name) my position is.
Repeat. Then wait for USCG to respond.

The Red Distress button

Digital Selective Calling. Automatically transmits a distress call with your GPS position.

Use only when unable to make radio call.

Pan Pan

Distress information broadcast made by the CG to inform other boaters of a distress.

Securitee Securitee
Safety information broadcast made by the CG.

Once you have established communication
with the other station, switch to another channel.

Ch-13

Vessel to Vessel Navigational safety.
Also to call draw or lift bridges.

Ch-16

Distress, safety and calling.
Calling the CG, other vessels and marinas, etc.

To make safety or security calls.

To make distress/mayday calls.

Channels 21, 22, 23

USCG working channels.

Channels 6, 9

Vessel to Vessel working channels.

Channels 68, 69, 71, 72, 78A

Marina and Vessel working.

Weather channels 1 through 7

National Weather Service
continuous weather broadcast.

Find the channel for your area or the one that broadcasts the information that pertains to you. Each area will have a different channel.

U.S. Customs

(202) 354-1000

All vessels required to clear in person at designated customs stations.

www.cpb.com *(Pleasure boat requirements by location)*

What you need:

- Vessel document or state registration. If new purchase and document not issued yet. *(Traveling letter from title company)*
- Annual decal number. www.dtops.cbp.dhs.gov
- Valid passport for each person onboard.

If not the owner of the vessel:

- Letter of authorization from boat owner.
- Vessel delivery, not full-time captain.
- Commercial entry form 1300 (02/02)

Items to consider:

- Alcohol . . . . 1 Liter/person over 21. (33.8 fl oz)
- Tobacco . . . . 200 Cigarettes/100 Cigars.
- Fruits and Vegetables. *(Recommend not having any onboard.)*
- Firearms. *(Must register)*
- Any prohibited/restricted items. Check cbp.com
- Cash of ten thousand dollars or more.

Note: For foreign built vessels. Recommended -Proof of initial import duty paid.

cbp.com (site navigation)

Click on: Travel, Pleasure Boats & Private Flyers

For locations by state and area click on: Pleasure Boat Requirements by Location

CBP Roam. Customs clearance App. Highly recommended

Canadian Customs

All vessels required to clear in person at designated customs stations.

Reporting and locations (888-226-7277)

What you need:

Vessel document or state registration.
(Traveling letter)
Valid Passport for each person onboard.

Items to consider:

Alcohol . . . . . . 1.5 Liters Wine or
. . . 1.14 Liters Alcohol or
. . . 8.5 Liters Beer/Ale

Tobacco . . . . . . 200 Cigaretts/50 Cigars
. . . . 200 Grams Tobacco
. . . . 200 Tobacco sticks

Fruits and Vegetables. *(Recommend not having any onboard)*

Firearms. *(Prohibited, must surrender)*

Any prohibited/restricted items.

Cash of ten thousand dollars or more.

Mexico (and all other countries)

When arriving Mexico from the US:

Temporary Import Permit. (Must have)
Crew List
Vessel Document or Registration
Valid Passport for each person on board
Fishing license for each person on board
If fishing gear is onboard
Mexico liability insurance

When arriving from another country

Or going to another country.
Zarpe/crew list from departing country
Valid Passports
Temporary Import Permit (if going into Mexico)
Mexico liability insurance

Mexico paperwork services

Cesar Ramirez
Paperman Ensenada
paperman.ensenada@gmail.com
52-646-223-9941

PAPERMAN
ENSENADA

CÉSAR RAMÍREZ
+52 (646) 223 9941 • +52 (663) 112 2060 paperman.ensenada@gmail.com

SERVICES

- *Arrival and departure paperwork (Harbor Master, Immigration and Customs)*
- *Temporary Importation (TIP)*
- *TIP Cancellation*
- *Skipper services*
- *Boat terrestrial & maritime transportation*
- *Yacht cleaning & detailing at the Marina*
- *Local transportation service, Mexico-USA and vice versa*
- *Marina slip reservations*
- *Info/reservation of yacht transportation service on freighter on the east and west coast*
- *Fuel delivery service at Hutchison Ports ECV Marina*

Anchoring Tips

Select a position. Suitable depth and bottom. Use three to five times the depth for amount of chain. Make note of tide range. Allow plenty of room for you and your neighboring vessels to swing.

Prepare the anchor for dropping. Approach position into the wind and current.

Drop the anchor. Allow enough chain on the bottom before allowing the vessel to drift back.

Let vessel sit back on scope. Use more scope if vessel continues to drift. Monitor vessel's drift and swing.

Med Mooring
Anchor drop distance off dock needs to allow for at least a 3 to 1 scope. Depth times three.

Set a stern line off Port and Starboard side and cross tie to the dock.

When in a cross wind, set anchor to the upwind side.

It will be necessary to have crew manning the anchor and the stern lines at the same time.

Shore Tied
The shore tie is used when relatively deep water runs close to shore and/or space is limited requiring vessels to anchor stern to the shore.

- A shore line of at least 250 feet is needed
- Select a suitable tree, rock, ect. to tie line to
- Pay close attention to tide range
- Use Tender to run the line to the beach

Quick Check Anchoring

- ✓ Prepare anchor for dropping
- ✓ Back anchor out of housing
- ✓ Select position and depth
- ✓ Bow into the wind and current
- ✓ Hold boat in position while dropping anchor
- ✓ Let out suitable scope. 3–5 to one
- ✓ Let boat set back on chain
- ✓ Check for drift

Using The Ballard Locks

What you need:

- Bow and Stern Line.
- Adequate Fenders.
- (2) 50 ft lines for Large Locks.

How to Proceed:

Small Locks: First come, First serve.
Large Locks: Larger boats enter first.
(Smaller vessels expect to raft to larger vessels)
Westbound: Expect Starboard side.
Eastbound: Expect Port side.
(Be prepared to switch sides if directed)
Red and Green traffic style lights will indicate when to proceed.

What To Do:

Small Lock: Secure the eye of your line to the bow and stern cleats. Run the line around the numbered bollard on top of the sleave that lines the sides of the small locks and then back to the boat. The boat will then move with the sleave as the water level rises or falls.

Large Lock: Hand the Lock Attendent the eye of your line. Secure the line to your bow and stern cleats. Standby your bow and stern line. Adjust lines in or out with the rise or fall of the Vessel in the locks. Keep tension on the lines.

Do not release lines until directed by the Lock Attendants.

Pay attention to, and follow the direction of the Lock Attendant

Maneuvering Tips

- ☑ Keep control inputs slow and deliberate.
- ☑ Use one hand and one control at a time.
  This will help keep you under control.
- ☑ If you feel the situation getting out of control,
  STOP. Take your hands off the controls.
  Take a breath. Do Not Panic. Stay calm.
  THIS IS EXTREMELY IMPORTANT!
- ☑ Always be aware of you vessel's drift in all axes.
  Pick reference points to check your drift.
- ☑ Your vessel is a natural weather vane. It is always easier to back into the wind and current. The bow will follow.
- ☑ Always lead your turns. Your turning radius will always be bigger than expected.
- ☑ Always do your maneuvering upwind and up current.
- ☑ Always be aware of which direction the wind and current will set you.
- ☑ Be aware of your momentum. Let it work for you.
- ☑ Let the wind/current work for you, not against you.
- ☑ Make a plan and ensure that lines and fenders are set before you make your approach.

Passes and Narrows

Many passes and narrows are extremely hazardous during max current.

Plan your arrival and transit as close to slack water as possible.

Seakeeping

Hull design, vessel weight, rudder size and speed are all factors in how each boat will handle different sea conditions.

When possible avoid heading directly into sea swell and wind chop.

Put seas off to the port or starboard of the bow. Do not put seas directly on the beam. Run as close to down swell and chop as possible.

When in uncomfortable conditions, play with heading and speed to find the best ride possible. Most boats are more stable at higher speeds, so you may find that actually increasing speed helps your ride.

Distances

The Big Picture

Juneau, AK to Seattle, WA	902
Seattle, WA to San Diego, CA	1300
San Diego, CA to Panama	2870
Panama to Fort Lauderdale, FL	1250
Fort Lauderdale, FL to Newport, RI	1200
Via the Intracoastal Waterway	1400

East Pacific Ocean

Cape Flattery, WA to Honolulu, HI	2260
Cape Flattery, WA to Dutch Harbor, AK	1595
Cape Flattery, WA to Kodiak, AK	1130
San Francisco, CA to Honolulu, HI	2060
Newport Beach, CA to Honolulu, HI	2220

Gulf of Alaska

Juneau, AK to Elfin Cove (Cape Spencer), AK	100
Cape Spencer, AK to Yakutat, AK	132
Yakutat, AK to Seward, AK	307
Seward, AK to Homer, AK	160
Homer, AK to Anchorage, AK	140
Kodiak, AK to Anchorage, AK	205
Cape Spencer, AK to Hinchinbrook, AK	338
Hinchinbrook, AK to Valdez, AK	65
Hinchinbrook, AK to Whittier, AK	70

SE Alaska

Juneau, AK to Seattle, WA	902
Sitka, AK to Petersburg, AK	160
Juneau, AK to Petersburg, AK	114
Petersburg, AK to Ketchikan, AK	117
Ketchikan, AK to Prince Rupert, AK	90
Prince Rupert, AK to Hartly Bay, BC	80
Hartly Bay, BC to Bella Bella, BC	110
Bella Bella, BC to Port Hardy, BC	100
Port Hardy, BC to Alert Bay, BC	30
Alert Bay, BC to Campbell River, BC	80
Campbell River, BC to Nanaimo, BC	70
Nanaimo, BC to Seattle, WA	120
Nanaimo, BC to Friday Harbor, WA	57
Shilshole, WA to Friday Harbor, WA	60
Friday Harbor, WA to Vancouver, BC	60

West Coast

Seattle, WA to San Diego, CA	1300
Seattle, WA to Newport Beach, CA	1250
Seattle, WA to San Francisco, CA	850
Seattle, WA to Portland, OR	350
San Diego, CA to Cabo, MX	740
Cabo, MX to Canal	2410
San Diego, CA to Canal	2870
Canal to Fort Lauderdale, FL	1250
Canal to Isla Mujeres	830
Seattle, WA to Neah Bay, WA	100
Neah Bay to Westport, WA	100
Westport, WA to Columbia River, WA	40
Columbia River to Newport, OR	100
Newport, OR to Coos Bay, OR	75
Coos Bay, OR to Crescent City, CA	100
Crescent City, CA to Eureka, CA	65
Eureka, CA to Bodega Bay, CA	175
Bodega Bay, CA to San Francisco, CA	65
Bodega Bay to Monterey, CA	125
San Francisco, CA to Monterey, CA	90
Monterey, CA to Morro Bay, CA	100
Morro Bay, CA to Santa Barbara, CA	100
Santa Barbara, CA to Newport Beach, CA	100
Newport Beach, CA to San Diego, CA	75

Baja/Sea of Cortez

San Diego, CA to Ensenada, MX	65
Ensenada, MX to Turtle Bay, MX	275
Turtle Bay, MX to Cabo, MX	400
Cabo, MX to LaPaz, MX	150
Cabo, MX to Puerto Vallarta, MX	300
Cabo, MX to Mazatlan, MX	190
Mazatlan, MX to La Paz, MX	240
Mazatlan, MX to Puerto Vallarta, MX	180
Puerto Vallarta, MX to Bara, MX	130

Mexico

Barra, MX to Chiapas MX.	775
Cabo, MX to Bara, MX	375
Bara, MX to Ixtapa, MX	210
Ixtapa, MX to Acapulco, MX	120
Acapulco, MX o to Huatulco MX	240

Distances

Central America

Huatulco, MX to Barillas, El Salvador	500
Huatulco, MX to Puerto Madero, MX	223
Huatulco MX to Chiapas MX	223
Huatulco MX to Puerto Quetzal	335
Huatulco, MX to Papa Gayo, CR	685
Huatulco, MX to Los Suenos	775
Puerto Madero, MX to Los Barrillas, MX	284
Los Suenos to Puerto Quetzal	454
Chiapas MX to Puerto Quetzal	115
Chiapas MX to Papa Gayo	470
Chiapas MX to Los Suenos	560
Puerto Quetzal to Puerto Madero	115
Puerto Quetzal to Huatulco	335
Puerto Quetzal to Papa Gayo	360
Puerto Quetzal to Los Suenos	454
Barillas to Puesta Del Sol, Nic	70
Barillas to Papa Gayo	230
Barillas to Los Suenos	345
Puesta Del Sol to Papa Gayo, CR	100
Papa Gayo to Los Suenos, CR	125
Papa Gayo to Panama	555
Los Suenos, CR to Golfito	150
Golfito to Panama	345
Los Suenos to Panama	450
Canal to Fort Lauderdale	1250
Colon (canal) to San Andreas	220
San Andreas to Roatan	430
Roatan to Isla Mujeres	310
San Andreas to Caymans	410
Caymans to Isla Mujeres	325
Isla Mujeres to Key West, FL	350
Key West, FL to Fort Lauderdale, FL	165

Canal to Fort Lauderdale 1250

Colon (canal) to San Andreas	220
Colon to Roatan	650
Colon to Caymans	630
Colon to Isla Mujeres	835
San Andreas to Roatan	430
Roatan to Isla Mujeres	310
San Andreas to Caymans	410
Caymans to Isla Mujeres	325
Isla Mujeres to Key West	350
Key West to Fort Lauderdale	165

Via Windward Passage

Fort Lauderdale, FL to Mathew Town	516
Mathew Town to Port Royal	247
Fort Lauderdale, FL to Port Royal	760
Port Royal to San Andreas	435
Port Royal to Colon	545

East Coast

New York City to Fort Lauderdale, FL	1200
Newport, RI to Fort Lauderdale, FL	1300
New York City to Cape May	135
Cape May to Norfolk, VA	170

East Coast Intracoastal Waterway

Norfolk, VA Statute Mile Marker	000
Great Bridge Locks	11.5
Great Bridge (Atlantic Yacht Basin)	012
Coinjock, NC	050
Belhaven, NC	135
Oriental, NC	180
Moorehead City, NC (Cape Lookout)	205
Swansboro, NC	230
Wrightsville, NC	285
Wilmington, NC (Cape Fear)	295
Southport, NC (Cape Fear)	310
N. Myrtle Beach, SC (Myrtle Beach YC)	345
Myrtle Beach, SC (Barefoot Resort)	355
Georgetown, SC	405
Charleston, SC (Entrance)	470
Beaufort, SC	535
Hilton Head, SC (Harbor Town)	565
Savannah River (Savannah + 5 mi)	575
Thunderbolt, GA	585
Lanier Is, GA (Golden Isle Marina)	675
Fernandina Beach, FL (Entrance)	715
Jacksonville, FL (St. Johns River Entrance)	740
St. Augustine, FL (Entrance)	780
Daytona Beach, FL	830
Ponce de Leon Inlet	840
Titusville	880
Melbourne, FL	920

Distances

Vero Beach	950
Fort Pierce (Entrance)	965
St. Lucia (Inlet)	990
Jupiter (Inlet)	1005
Palm Beach, FL (Entrance)	1018
Fort Lauderdale, FL (Entrance)	1065
Miami, FL (Entrance)	1090

Chesapeake Bay

Norfolk, VA to Solomons Is, MD	100
Solomons Island to Annapolis, MD	44
Annapolis, MD to Baltimore, MD	24
Baltimore, MD to Cape May, NJ via C&D Canal	110

East Coast Coastal Routes

Fort Lauderdale, FL to New York City direct	1055
Newport, RI to New York City	140
New York City to Atlantic City, NJ	100
Atlantic City, NJ to Cape May, NJ	34
Cape May, NJ to Chesapeake Bay Entrance	130
To Norfolk, VA	35
Chesapeake Bay to Cape Lookout, NC	190
To Moorehead City, NC	22
Cape Lookout, NC to Cape Fear, NC	80
To Southport, NC	33
To Wilmington, NC	55
Cape Fear to Charleston, SC	110
To Charleston	21
Charleston, SC to Savannah River	70
To Savannah, GA	22
Savannah River to Jacksonville, FL	100
Jacksonville, FL to Fort Lauderdale, FL	271
Fort Lauderdale, FL to Miami, FL	19
To Miami, FL	6
Fort Lauderdale, FL to Key West, FL	167

Bahamas

Fort Lauderdale, FL to Bimini	50
Fort Lauderdale, FL to West End	70
Fort Lauderdale, FL to Lucaya	85
Palm Beach, FL to Lucaya	78
Lucaya to Nassau (Paradise Island)	112
Bimini to Chub Cay	87
Chub Cay to Nassau (Paradise Island)	38
Nassau to Harbor Island (Dunmore)	55

Bermuda

Hamilton to Newport RI	640
Hamilton to Ambrose Light	660
To NYC	20
Hamilton to Chesapeake Bay Entrance	628
To Norfolk	35
Hamilton to Charleston	786
Hamilton to Fort Lauderdale, FL	912

Carribean

Fort Lauderdale, FL to Mathew Town	516
Mathew Town to San Juan PR	452
San Juan, PR to St. Thomas	70
San Juan, PR to St. Croix (Christianstad)	100
St. Thomas to St. Martin (Simpson Bay)	108
St. Martin to St. Barts (Gustavia)	17
St Barts to Antigua (St. Johns Hbr)	90
Antigua to Martinique (Fort-de-France)	171
Martinique to St. Lucia (Rodney Bay)	33
St. Lucia to St. Vincent (Kingstown)	64
St. Lucia to Barbados (Bridgetown)	120
St. Vincent to Barbados (Bridgetown)	93
St. Vincent to Grenada (St. Georges)	75
Grenada to Trinidad (Chaguaramas)	87
Trinidad to Bonaire	415
Bonaire to Curacao (Willemstad)	40
Curacao to Aruba (Oranjstad)	70
Aruba to Colon Panama	624

See Articles Written by Chris Couch at CompassHeadings.com

Featured in Northwest Yachting Magazine:

The Gold Standard
September 2017

Lost Horizon
July 2017

The Trip
June 2017

Unlocking the Locks
March 2017

The Overnight Passages
January 2017

Anchor Management
November 2016

Weather of the West
October 2016

Special Delivery
September 2016

The Go / No Go Decision
August 2016

...these and more

Yacht Reviews & Testimonials:

54' Ocean Alexander
70E Ocean Alexander
78' Ocean Alexander

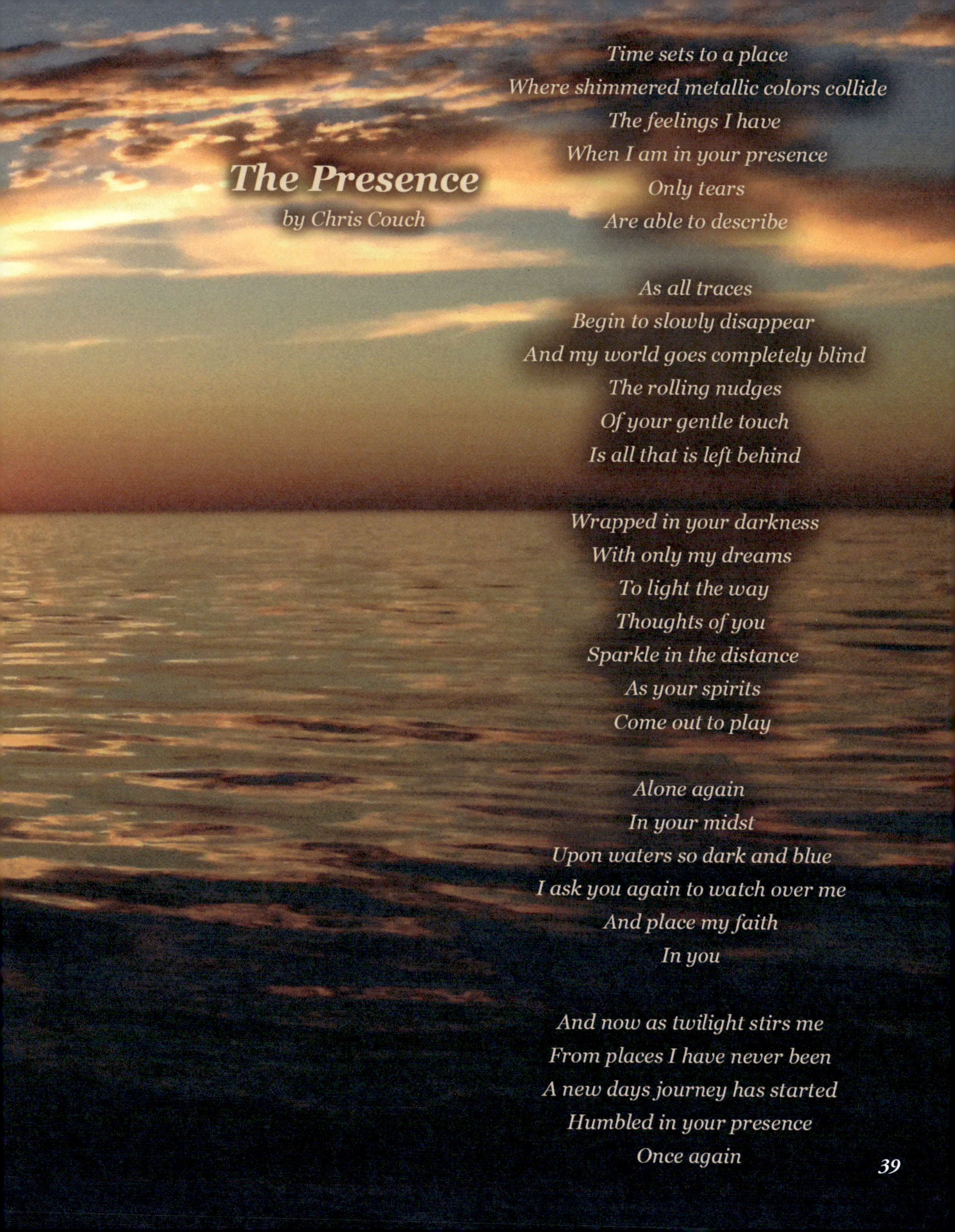

The Presence

by Chris Couch

Time sets to a place
Where shimmered metallic colors collide
The feelings I have
When I am in your presence
Only tears
Are able to describe

As all traces
Begin to slowly disappear
And my world goes completely blind
The rolling nudges
Of your gentle touch
Is all that is left behind

Wrapped in your darkness
With only my dreams
To light the way
Thoughts of you
Sparkle in the distance
As your spirits
Come out to play

Alone again
In your midst
Upon waters so dark and blue
I ask you again to watch over me
And place my faith
In you

And now as twilight stirs me
From places I have never been
A new days journey has started
Humbled in your presence
Once again

Alaska / Canada

Elfin Cove
Fuel 907-239-2208
Limited moorage on gov't. docks.

Juneau
Port of Juneau 907-586-5720

Auke Bay ..... Harbor Master 907-789-0819
Fishermans Bend 907-789-7312
Deharts Marina 907-789-4225
Seadrome Travel Center 907-463-8811

Petersburg 907-772-4688
Call ahead. Moorage limited.

Ketchikan 907-228-5637
City float, just north of downtown.
New docks with power and water.

Between berths 3 and 4
Doyons Landing, Large Yacht moorage 907-225-5155
...... moorage@mistyfjordsair.com
Petro Marine. Will deliver to
Doyons Landing. 907-225-2106

Ketchikan Moorage
1285 Tongass, next to Berth 4 907-225-8285
Donna
3 long side ties

Prince Rupert
PRYC 250-624-4317
Outside end tie. Clear Customs after tieing up.
Canada Customs 888-226-7277
Fuel dock next door.

Hartley Bay .... fuel-limited moorage

Klemtu .... (hole in the wall) (fuel)

Bella Bella
Shearwater Marina (fuel) Ch-66A .. 250-957-2666

Butedale Abandoned Cannery. Decent dock for overnight.

Port Hardy (fuel)

Quarterdeck Marine 250-949-6550
Limited moorage

Customs Canada 888-226-7277

Alert Bay (fuel) 250-974-5213

Port McNiel. North Island Marina 250-956-4044
Fuel. Good services.

Cambell River
April Point 250-285-2222
Discovery Harbor 250-287-2614
Ch-66a

Cortez Island
Gorge Harbour Marina 50-935-6433
moorage@gorgeharbourmarina.com ... Ch-66A

Yaculta Rapids .... Transit near slack

Dent Island 250-203-2553

Big Bay. Stuart Is. Comm. Docks 250-202-3625

Sonora Resort (Yaculta Rapids) 604-233-0460

Toba Wildernest 250-830-2269
fishingwildernest.com

Lund
Hotel Dock 604-483-3187
Just north of fuel dock. (exposed)

Pender Harbor
Garden Bay Marina 604-883-2674
gardenbaypub.com gbhm@dccnet.com
35 amp only. Lots of dock space.

Pilot House Marin................. 604-883-2479
thepilothousemarina.com Ch-66A 50 amp.

Princess Louisa Inlet/Chatterbox Falls.
Gov't. dock 55-ft limit.
Shore anchoring only.
Malibu Rapids.......... Reference Pt. Atkinson
Canadian Tide Tables, Volume 5.
Add 1 hour to predicted times.
Transit at slack.

Nanaimo, Port of 250-754-5053

Ganges 250-537-5242

Montegue Harbor 250-539-5733

Bedwell Harbor (Poets Cove) 250-629-3212

Cameron Island 250-754-5053
Ch-67

Port of Sidney Ch-66A.............. 250-655-3711
Van Isle Marina (fuel/Customs) 250-656-1138
Philbrooks boat yard 250-656-1157

Raven Marine 250-655-3934
Anthony..Marine Repair/Management
Highly recommended.

Vancouver
Coal Hbr 604-681-2628
Taxi 871-1111

Victoria 250-383-8326
Ch-66A
Customs 888-226-7277
ABC Taxi. Jas. 250-744-0462
Coho 250-386-2202

Wrangell Narrows
Current flows north on the flood.

Washington

Bellingham Squalicum Harbor 360-676-2542

Anacortes
Cap Sante 360-293-0694

Roche Harbor 360-378-2155

Friday Harbor 360-378-2688
Customs 360-378-2080
Friday Harbor Taxi 360-298-4434

Rosario Resort 360-376-2222

La Conner
Swinomish Channel Lodge 360-466-3101

Port Townsend
Point Hudson 360-385-2828

Seattle Bell Street 206-615-3952

Elliott Bay Marina 206-285-4817

Shilshole Bay Marina 206-728-3006
Fuel Dock 206-783-7555

Poulsbo 360-779-3505

Gig Harbor Arrabellas 253-851-1793

Bellevue Marina 425-452-6881

Port Angeles 360-457-4505
Customs 360-457-4311

Neah Bay
Marina: 360-645-3015
Fuel: 645-2749
Big Salmon: 645-2374 Ch-68

Westport 360-268-9665
Fuel: 268-0076

Oregon

Astoria 503-325-8279
West Basin 503-791-7730
Englund Marine 503-325-4341

Port of Rainier – 1 mi. east of bridge across from Longview.
Good overnight moorage. No power.

Kalama 360-673-2325
Gate: 1121

Portland
Tomahawk Bay Marina 503-283-2444

Newport, OR
Southbeach: 541-867-3321
J Dock 50 amp
Fuel: 867-7644
After-hours fuel: 961-9037
Taxi: 265-9552
Mechanic: 265-4224
Ben's Diving 961-8471
US Customs 541-265-6456
Note: Call 24 hrs in advance.

Umpqua
Salmon Harbor 541-271-3407

Coos Bay
Charleston 541-888-2548
Taxi 888-4343
Mechanic 888-2623

Brookings 541-469-2218
200 ft of transient space with 50 Amp across from CG Station. Entrance is small over 60 ft not recommended.

California

Crescent City 707-464-6174
Fuel 707-465-4200
(call 30 minutes prior to arrival)
Taxi 707-464-6030
New Marina is complete. Transient moorage is long side tie. On the left as you enter the inner harbor. 50 amp power. Fuel pier is on the right just before the inner harbor.

Eureka
Woodly Island 707-443-0801
Englund Marine 444-9266
Taxi 445-1100

Fort Bragg 707-964-4719
Entrance is very small. Moorage is very limited. Over 50 ft not recommended.

Bodega Bay
Spud Point 707-875-3535 Ch-68

San Francisco
St Francis YC 415-850-1788
P-38 415-279-5562
South Beach 495-4911
SF Marina 292-2013

Sausalito
Spoonmaker 415-331-5550

Oakland
Grand Marina 510-865-1200
Port of Oakland 834-4591
Marina Village 521-0905
Emerycove 428-0505

Halfmoon Bay 650-726-5727 Ch-74

Monterey 831-646-3950
Breakwater Cove 647-9402
Diane 373-7857
Monterey Airbus 831-373-7777
Monterey Taxi 646-1234

Morrow Bay 805-772-6254
Yacht Club 772-3981

Santa Barbara 805-564-5530
Fuel 962-7186
Taxi 966-6161
Ocean Aire Electronics 805-962-9385
Marine Services 805-698-9614

Ventura
Harbor Village 805-644-0169
West Marina 644-8266

Marina Del Rey
Burton Chace Park............... 310-305-9595
Fuel 823-2444

Newport Beach
Ardell 949-642-5735
Bay Club 645-5000
Fuel (Gary Hill) 675-0740

Catalina
Avalon 310-510-0535
Two Harbors 510-2683

Dana Point 949-496-6137

San Diego
Kona Kai Marina 619-224-7547
jfischetti@konakaimarina.com
Shelter Island Marina 619-223-0301
marina@islandpalms.com
Seanet Electronics 619-222-3407

Marina Cortez Harbor Island
Mark Hallmark. 619-246-6676

Point Loma Sportfishing
Temp big boat Moorage 619-223-1627
John Klein 619-247-8297
Qualifier105klein@yahoo.com

San Diego Electronics
Toby 619-520-5028
Furuno, best there is.

Mission Bay
Driscoll 223-5191
Cabrillo 619-297-6222

Hawaii

Oahu, West side
Ko Olina Marina 808-679-1050
Koolinamarina.com

Mexico

Mexicoboating .com

Ensenada
Hotel Coral 866-302-0066
Cruiseport Village
Ecpvmarina.com
Marina.ensenada.net.mx

Turtle Bay – Old steel pier. Med Moor. Fuel is cash only.

Cabo
Marina Cabo San Lucas. . CSL@IGYMarinas.com
Fuel, large marina.
Mexico checkin and checkout
caboportagent.com daniel@caboportagent.com
Documents needed: Vessel Document, Import Cert.
Mexico vessel insurance, Passports.

Marina del Rey, Cabo San Lucas
bhopper@marinadelreycabo.com
Fuel dock and marina. Just on the left, as you enter.

Puerto Los Cabos, San Jose del Cabo
marinainfo@puertoloscabos.com
52 624 105 6028

Cabo Diesel Supply
Calle Ejido

San Jose Del Cabo
Puerto Los Cabos Marina . . . . puertoloscabos.com

Cabo Los Frailes – Good anchorage south side.

Cabo Pulmo Marine Park – Only hard coral reef in Sea of Cortez. No Anchoring. Need Guide.

Ensenada de los Muertos . . . . Good anchorage. Just south of Punta Perico.

LaPaz
Marinacostabaja.com
marinalapaz@bajavillas.com
Costabajainfo@bellportgroup.com
Marinapalmira.com
877-217-1513
011-52-612-125-3959

Beyond the Horizon

by Chris Couch

Where is the place
You go
When you dream?

Do your hopes
Take you past
The horizon
So very far?

Which direction
Do you face?
When you wish upon
Your favorite star

Who is the
Dreamer
From whom
You cannot hide?
How will you find
The believer
Still lost
Inside?

When you
Face yourself
In the mirror
What will you
Say?
Do you see the light
Within?
That will help
Show you
The way

Do you have the
Courage
To do?
And not just try
In what direction
Does your true
Faith and belief
Lie?

Take that first step
To make it
Come true
The only thing
That stands in your way
Is you

Follow your dreams
With the setting
Sun
To that place
That lies
Beyond the Horizon

Rancho Viejo (1 block up from waterfront just before marina lapaz)

Sea of Cortez (west side)

Starting North of LaPaz

Isla Espiritu Santos . . . . Many great beachs/anchorages

Isla Partida Anchorages
Partida Cove
El Cardonal
Ensenada Grande

Los Islotes . . . . Swim with the Seals.

Isla San Francisco – 45 nm North of LaPaz, Amazing clear water. Trail along ridge. Fantastic views.

Puerto Escondido . . . . Very sheltered

Loreto – Small boat harbor. Dinghy dock. No anchorage. Alaska Airlines

Isla Carmen – Across from Loreto
Balandra Cove. Good Anchorage.

San Juanico Good anchorage. Very pretty harbor.

Copper Canyon Train – Topolobampo. Stay in marked channel. Train leaves from Los Mochis. Anchor just outside channel off Mexican Military Base.

Mazatlan-LaPaz Ferry –ferrys@matur.com.mx

Mazatlan
Elcid.com U.S. 866-796-5571
23-16.0N 106-28.0W
5.4 mi N. of Main entrance.
marinaelcidmazatlan@elcid.com.mx
52 669 916-3468

Puerto Vallarta
Marina Vallarta 011-52-322-221-0275
Paradise Village . . . . . marina@paradisevillage.com

Barra de Navidad
Grand Isla Navidad Resort. . . . . . . islanavidad.com
harbormaster@islaresort.com.mx
Fuel dock just beyond marina.

Saul Alverez Dock Master
. . . . secundinoa@hotmail.com
. . . . archie_107@hotmail.com
French pastry Panga El Harna Frances
delivered each morning
VHF-22
christeme@hotmail.com
Water Taxi Ch-23

Manzanillo – Las Hadas Resort. NE of Punta Santiago
Ixtapa . . . . ixtapa@jarestate.com
Lots of moorage and fuel
Acapulco . . . . Yates@acabpu.com.mx

Ixtapa . . . . Marinaixtapa.com.mx
Ixtapa@jarestate.com
Lots of big boat Moorage/fuel
Shoaling reported at entrance

Huatulco Marina Chahue. 15-44.9N. 096-06.7W
Marinachahuereserv@fonatur.gob.mx

Puerto Chiapas (old Puerto Madero)
memogarcia@marinachiapas.com.mx
Checkin/checkout services
Several long endties to 100 ft plus.
50 amp only. Relatively new Marina with Restaurant
City of Tapachula nearby with airport

Guatamala

Puerto Quetzal
13-54.5N 090-45.8W
Marina Pez Vela marinaguarte.com
customerservice@marinaguarte.com
Agent Miguel Oscar: VHF-10
tramsmaritimos@intellnet.com
90 ft. end tie with 100 and 50 amp power.

El Salvador

Bahia Jiquillisco
Las Barillas
Barillasmarina.com
info@barillasmarina.com
13-07N 88-25.2W
Pilot pickup point. Ch-16
Mooring buoys. Fuel dock.
Marina will notify officials.
9 mi up river to marina.
550 nm to Huatulco.

Costa del Sol
Paradise Fishing Lodge
Paradisefishinglodge.net
info@paradisefishinglodge.net

Nicaragua

Marina Puesta del Sol.com
. . . . .mpuestadelsol@yahoo.com
5 mi off 12-35.6N 087-25.5W
Sea Buoy 12-36.5N 087-22.4W
Just south of Golfo de Fonseca, 2 hrs. from Managua.
011-505-880-0019

70 nm to Las Barillas
100 nm to Papagayo
250 nm to Los Suenos

Costa Rica

Marina Papagayo . . . . 10-38.25 N. 85-39.25 W
NW side of Bahia de Culebra
marinapapagayo.com
Agent: Ernesto . . . . paramares@hotmail.com
Liberia. Nearest international airport

Los Suenos . . . . 9-38 N 84-40 W
Across from Cabo Blanco
info@lossuenosresort.com
866-865-9759
011-506-637-8886
Agent: Ernesto. . . . . paramares@hotmail.com
. . . . 24 hr notice.
Electrical: . . . . Preston Wright
. . . . marinemetro@racsa.co.cr
Marine Power . . . . .Luis Madrigal E.
. . . . . . lmadrigal@tractomotriz.com
2 hr cab ride from San Jose.
Dental: Dr. Dario Chaves O.
dariochaveso@hotmail.com
Town of Jaco.

Quepos 33 miles SE of Los Suenos
. . . . 9-25 N 84-10 W
. . Marinapezvela.com 866-739-8352

Golfito: Bananabaymarina.com
08-37N 83-09W
011-506-775-0838

Panama

There are several new marinas on the Panama City Side. Have the agent check for availability.

Marina Flemenco Ch-10
Flamenco Marina: fuerteamador.com
8-54.37 N 79-31.3 W
East side of Flemenco Is.
marina@fuerteamador.com

Colon . . . . (Cristobol Control ch-12)
ShelterBayMarina.com (Ch-74)
info@shelterbaymarina.com, 9-22.23N 079-56.864W
Keep "SB" green buoys to left and the breakwater to the right.

Cayman Islands

Georgtown Barcadere Marina. . . . .
. . . . enquiries@barcadere.com

New Jersey

Atlantic City 609-441-2000
Trumpmarina.com

Cape May 609-884-2400
Southjerseymarina.com

Delaware

C & D Canal, Bear, DE 302-836-1800
ummitnorthmarina.com

Maryland

Baltimore 410-752-1122
Harborview Marina

Anapolis 410-268-0660
Anapoliscitymarina.com

Solomons
Solomons Yachting Center 410-326-2401
Harborislandmarina.biz 410-326-3441
Zahnisers Yacht Center 410-326-2166

Virginia

Norfolk ICW mile 0
Watersidemarina.com 757-625-3625
Portsmouth Tide Water Marina
Tyamarina.com 757-393-2525

Great Bridge
Atlantic Yacht Basin 757-482-2141

North Carolina

Coinjock
Midway Marina 252-453-3625
Coinjock Marina 252-453-3271

Belhaven
River Forest Marina 252-943-2151

Oriental
Oriental Harbor Marina 252-249-3783
Moorhead City. 252-726-6862
Moorheadcityyachtbasin.com

Swansboro
Caspers Marina 910-326-4462

Wrightsville Beach
Wrightsville Beach Marina. 910-256-6666

Wilmington
Wilmington City Docks 910-520-6875

Southport
Southport Marina 910-457-5261

South Carolina

North Myrtle Beach (Little River)
Myrtlebeachyachtclub.com 843-249-5376
Barefootresortclub.com 843-390-7928

Georgetown
Harborwalk Marina. 843-546-4250

Isle of Palms
Isle of Palms Marina. 843-886-0209

Charleston
Charleston City Marina 843-723-5098
Megadock.us

Beaufort
Downtown Marina 843-524-4422
Hilton Head
Harbourtownyachtbasin.com. 843-671-2704

Georgia

Savannah
Westin Savannah Harbor Resort 912-201-2000

Thunderbolt
Bahiableumarina.com
912-354-2283

St. Simons Island
Golden Isles Marina
Gimarina.com 912-634-1128

Florida

Customs: Florida ICE 800-432-1216

Fernandina Beach
Fernandina Harbor Marina 904-491-2090

Jacksonville/Mayport
Morningstar Marina 904-246-8929

St. Augustine 904-825-1026
Staugustinemarina.com

Daytona Beach
Halifax Harbor Marina 386-671-3601

New Smyrna Beach
Riverview Hotel 386-428-5858

Titusville
Titusville Municiple Marina 321-383-5600

Fort Pierce
Fortpiercecitymarina.com 772-464-1245

Stuart
Sunset Bay Marina. 772-238-9225

Palm Beach
Rybovich Marina 561-844-1800

Fort Lauderdale
Los Olas Marina 954-828-7200
Bahia Mar 764-2233
P-66 728-3578

Harbour Towne 926-0300

Miami
Miami Beach 305-673-6000
Miami Bayside 579-6955
Dinner Key 579-6980

Key West
A & B Marina.com 305-294-2535
Galleon Resort and Marina 305-292-1292

Bahamas

Bimini
Biminibayresort.com
242-347-2900
Biminibiggame.com
232-347-3391

Grand Bahama (west end)
Oldbahamabay.com

Great Inagua (Mathew Town)
Mathew Town Dock
Leon Turnquest 242-339-1427
Harbor Master, Donna McPhee

Jamaica

Port Royal/Kingston
Morganshabour.com
876-967-8040

Mexico (Yucatan)

Isla Mujeres
Puertoislamujeres.com
Marina@puertoislamujeres.com
Fuel, moorage, clearance assistance.
North end of Laguna Makax.
Check in with Port Captain on 16

Honduras

Roatan (French Harbor)
Barefootcay.com
Info@barefootcay.com
Fuel, clearance assistance.
Enter through Brick Bay entrance.
Agent Beatman Ebanks
beatman@globalnet.hn
Check in with Port officials on 16 off Coxen Hole.

Roatan, Honduras (Brick Bay)
To San Andreas, Columbia
Deep water route:

16-19.8 N 086-28.6 W	Brick Bay Entrance
16-06.0 N 085-01.0 W	Cabo Camaron
16-00.0 N 084-16.0 W	Punta Patuca
15-36.0 N 083-01.0 W	Mosquito Shoal
15-15.3 N 082-18.6 W	Largate Reef
12-35.0 N 081-40.0 W	North End San Andreas
Stay outside reef	
12-32.6 N 081-40.4 W	San Andreas Entrance

Columbia

San Andreas
Nenes Marina. Fuel. Med moor
Agent Rene, Agencia Serrana
Hail Costa Guarda on 16 at entrance.
12-32.2N 081-41.2W
Announce arrival. Name, POB. Homeport.
Request permission to enter.
Agent. Julian Watson • agenportltda@hotmail.com

Cayman Islands

Grand Cayman
Caymanislandsyachtclub.com
19-22.84N 081-19.6W
Entrance Gap to Main Channel
North end of Govener's Harbor
Call Port Security with 10nm on 16.

Carribean

Useful Web Sites:
Marinas.com
Noonsite.com
Tripsailor.com

San Juan, Puerto Rico
San Juan Bay Marina
Sjmarina.com

St. Thomas
Yacht Haven Grand
YHGST@igymarinas.com
American Yacht Harbor Marina
Americanyachtharbor.com

Tortola (Road Town)
Village Cay Marina
Igy-villagecay.com

St. Croix (Christianstad)
Green Cay Marina
Tamarindreefhotel.com

St. Martin
Gateway Marina
gateway@caribserve.net

Simpson Bay Marina
Igy-simpsonbay.com

St. Barthelemy (Gustavia)
Marina St Barth Marine

Antigua
Antigua Yacht Club
Aycmarina.com
Jollyharbourmarina.com

Martinique (Forte-de-France)
Marin Yacht Harbour Marina
Marinyachtclub.com

St. Lucia
Igy-rodneybay.com

St. Vincent Grenadines
Bequia Marina (Admiralty Bay)
Noonsite.com

Barbados (Bridgetown)
Careenage Marina
Marinas.com

Grenada (St Georges)
Camper Nicholson
Cnportlouismarina.com

Trinidad (Chaguaramas)
Crews Inn Hotel and Yacht Center
Crewsinn.com

Bonaire
Harbour Village Marina
Harbourvillage.com

Curacao (Willemstad)
Kima Kalki Marina
Kimakalki@onenet.an

Aruba (Oranjstad)
Renasissancemarina.com

Panama Canal

Agents:

Matchship.com
davidmanrique@matchship.com
Pete Stevens
delfinomaritime@hotmail.com
Tina McBride
tinamcbride@hotmail.com

Marinas

Pacific:

Marina Flemenco 8-54.3 79-31.3
East side of Flemenco Is. Ch-10
marina@fuerteamador.com

OceanReefPanama.com
BuenaVentura.com.pa
VistaMarMarina.com

Atlantic:

Shelter Bay Marina 9-22.2 79-56.8
Keep SB green buoys to port.
Breakwater to Stb. Ch-74
info@shelterbaymarina.com

Prior to Arrival:

Agent Contact. Establish contact and provide all needed information. This should be done no later than one week prior to arrival.

Note: Information must be submitted to the Canal Authority no later than 96 hours prior to arrival.

Arrival

Pacific: Contact Flemenco Signal Ch-12
Atlantic: Contact Cristobal Control Ch-12
Note: Let them know you if you have an agent and your destination marina.

Agent Required Canal Information

Name of vessel
Panama canal authority ship ID number (if the vessel already has one)

Captain and Owner

Complete name of captain
Name and address of vessel's owner
E-mail address
Phone/fax/satphone

The Vessel

Radio call sign
Registration/official nr.
Flag
Home port
Ship builder/make
Where built
Date built
Cruise speed
Hull material
Hull color
LOA
Beam
Draft
Gross tons
Net tons
Vessel type
Motor/sail
Make and number of main engines
Engine serial numbers
Horsepower

Arrival

ETA Cristobal (Atlantic)
ETA Balboa (Pacific)
Last port before arriving the Canal
Next port after leaving the Canal

Requirements

Fuel and moorage needed on both sides.
Has the vessel transited the Canal before?
Do you have 4 extra crew that are capable of being line handlers?
Do you have the required 4 X 125 foot lines required for transit?
How much fuel do you consume at cruise in 24 hours?

State the type of transit desired:
Center lock, side wall, tug tied (recommended)

The Crew

List of crew onboard upon arrival.
List of crew onboard upon departure.
Name, DOB, place of birth, age, nationality, passport number.

Note: Persons arriving Panama embarking on the boat and then departing will require a special visa. Ensure that agent has complete list of persons arriving and who will be departing with the boat.

Chain of events

Upon arrival:
Boarded by customs, immigration, health.
Agent will schedule Canal Authority to board and
and measure the boat. Ship ID will be issued.
Once fee has been deposited, vessel will be placed on the schedule.

Note: Wiring funds to Canal Authority via the Agent in advance is recommended.

Deep Blue

by Chris Couch

Night fades, morning dawns
With lights gentle stream
I awaken to find myself
Living in the dream

Looking back at a life
I have left behind
Sailing towards a world
That waits on the other side

Breaking the bonds of an earthly grip
As I begin to float free
Enveloped in this state I have become
Alone on an open sea

Surrounded by so much
Having never felt so all alone
I stare upwards
At the stars of home

The strength of her hold
The power of her ties
Puts my head in my hands
And brings tears
To my eyes

Peering into the deep blue
Where it is only truth I see
Reflecting off this place

I know I should be

Notes

Be sure to check out the companion guide to the checklist:

Beyond the Horizon *by Chris Couch*

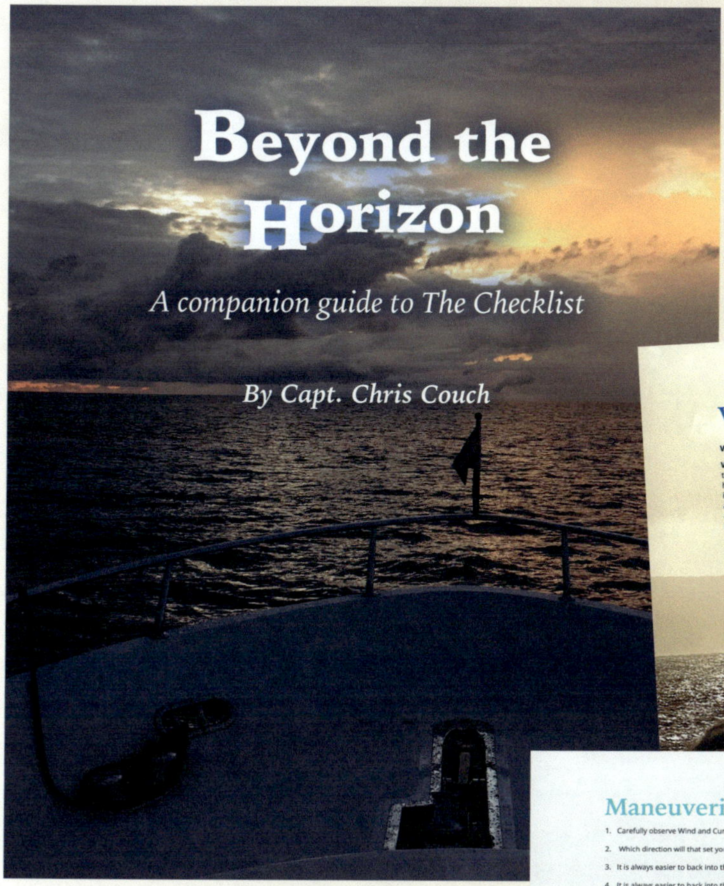

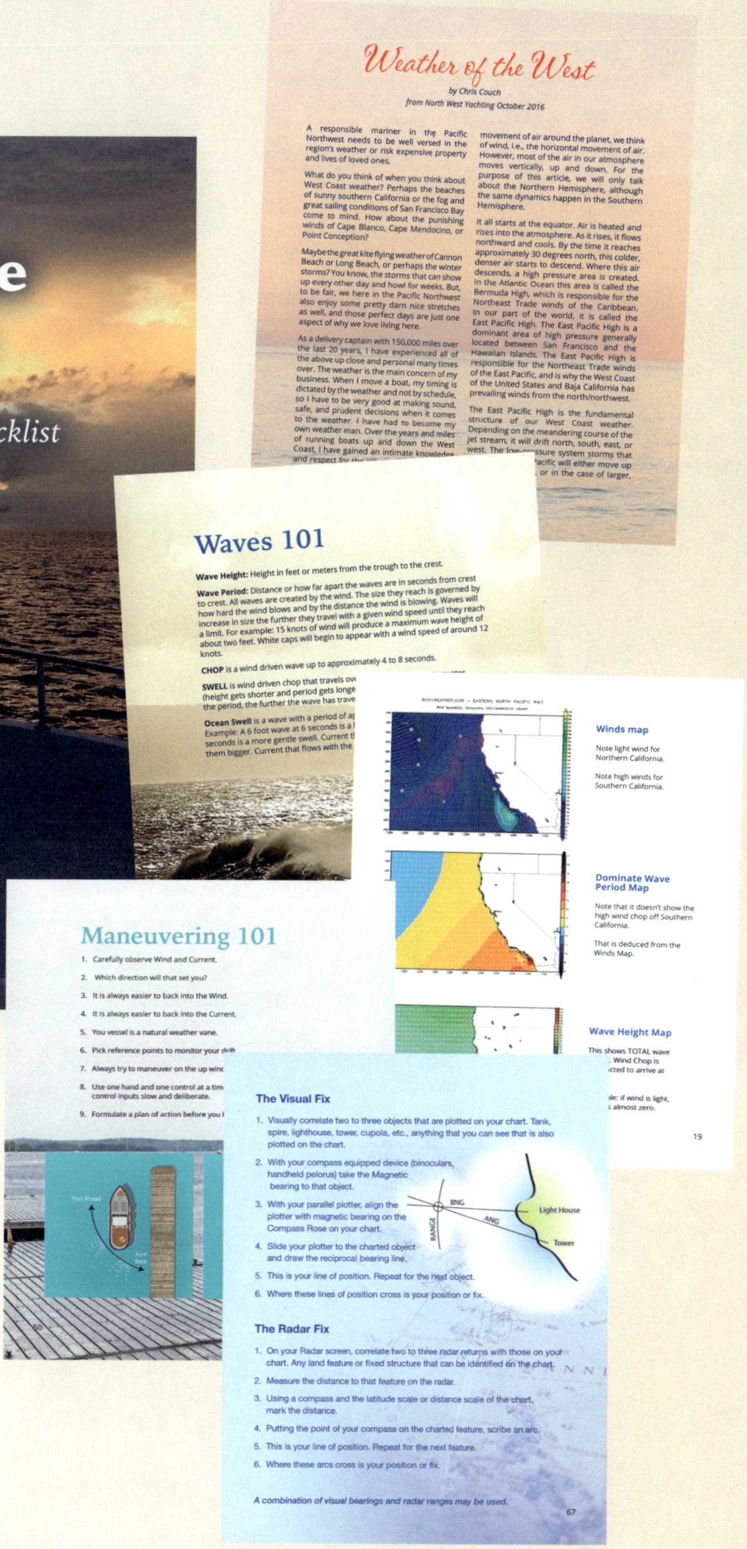

A companion guide to The Checklist.

Articles and in-depth guidance from Captain Chris Couch on Weather, Trip Planning, Maneuvering, Anchoring, Overnight Passages, Navigation.

Available on Amazon or at www.CompassHeadingsPublishing.com

www.CompassHeadings.com

Made in United States
Troutdale, OR
11/03/2026

56844460R00036